1 MONTH OF FREE READING

at

www.ForgottenBooks.com

By purchasing this book you are eligible for one month membership to ForgottenBooks.com, giving you unlimited access to our entire collection of over 700,000 titles via our web site and mobile apps.

To claim your free month visit: www.forgottenbooks.com/free271181

* Offer is valid for 45 days from date of purchase. Terms and conditions apply.

ISBN 978-0-428-97907-2
PIBN 10271181

This book is a reproduction of an important historical work. Forgotten Books uses state-of-the-art technology to digitally reconstruct the work, preserving the original format whilst repairing imperfections present in the aged copy. In rare cases, an imperfection in the original, such as a blemish or missing page, may be replicated in our edition. We do, however, repair the vast majority of imperfections successfully; any imperfections that remain are intentionally left to preserve the state of such historical works.

Forgotten Books is a registered trademark of FB &c Ltd.
Copyright © 2017 FB &c Ltd.
FB &c Ltd, Dalton House, 60 Windsor Avenue, London, SW19 2RR.
Company number 08720141. Registered in England and Wales.

For support please visit www.forgottenbooks.com

THE

ORIGIN, PROGRESS,

AND PRESENT STATE

OF

THE THAMES TUNNEL;

AND

THE ADVANTAGES LIKELY TO ACCRUE FROM IT,

BOTH TO THE PROPRIETORS AND TO THE PUBLIC.

SEVENTH EDITION.

LONDON:

EFFINGHAM WILSON, ROYAL EXCHANGE.

THE
ORIGIN, PROGRESS,
AND PRESENT STATE
OF
THE THAMES TUNNEL;
AND
THE ADVANTAGES LIKELY TO ACCRUE FROM IT,
BOTH TO THE PROPRIETORS AND TO THE PUBLIC.

SEVENTH EDITION.

LONDON:
EFFINGHAM WILSON, ROYAL EXCHANGE.

CONTENTS.

Introduction 5
Some account of former attempts to make a Tunnel under the Thames 7
Origin of the present Thames Tunnel Company 9
The Act of Parliament incorporating the Company 10
The purchasing ground, and commencing the work 11
The laying the foundation stone, the building the Tower and sinking it .... ib.
The lowering the Shield and placing it in its position 14
Description of the Shield 15
The materials and manner of working 16
Description of the Arches 17
Strata of the Shaft and Tunnel 18
The Expense and probable Revenue 20
The utility of the Tunnel to the Public, and the probability of advantage to the proprietors 21

INTRODUCTION.

For some time past a torrent of prejudice seems to have threatened to sweep before it every thing which has acquired existence, in the shape of a *Joint Stock Company;* and a sort of criminality is made to attach to every one connected with them; it may therefore be permitted to those who have given encouragement to *truly laudable Institutions of that nature,* to step forward and claim exemption from the general obloquy. At any other time and under ordinary circumstances, I should not think of intruding on public attention; but it seems a duty now, to step in between the public and the hasty censurers of so great an undertaking, and to endeavour to correct those false impressions that serve only to depreciate property, and thus to increase the difficulties which attend the establishment of all new Institutions. Acting from this feeling, I venture to solicit public attention to the real nature and value of

The Tunnel under the Thames,

which I boldly assert is calculated to do honour to our country; it ought to be considered as a great national work. It is unrivalled in modern times, and I believe without a parallel, as far as we have a knowledge of the works of the ancients.

In order to shew the real importance of this grand work, and to remove it entirely out of the pale of *Bubble Companies,* (which I pledge myself to do,) it will only be necessary to give a slight sketch of what has been done, and the manner of doing it; to state the advantages likely to arise from it to the City of London and the neighbouring counties, the great interest taken in it by engineers both civil and military, and the anxiety with which the progress of the work is watched by almost every man of science in Europe.

20th JULY, 1827.

Another edition of this little pamphlet being called for, it may be proper to say a few words as to the accident which has stopped the progress of the tunnel. The great interest the public have taken in this truly national work has made it a matter of notoriety, that in May the water of the Thames broke into the tunnel, between the termination of the brickwork and the shield, from causes which have been amply detailed in the newspapers. Measures were immediately set on foot for stopping the influx of water, which have, in a great measure, proved effectual; the water is so far reduced, that there is great probability the work will re-commence very soon, and proceed without further delay. It may, with great truth, be said, that no accident to any public work could have excited more sympathy and regret, and where success will be hailed with greater joy.

THE THAMES TUNNEL.

A GREAT variety of new and extensive undertakings have within these last two years attracted public attention; but it may be fairly said that none has appeared of greater importance than the *Tunnel under the Thames,* whether considered with the eye of an engineer and a friend to science, or with views of public utility and commercial importance.

The forming a spacious subterranean communication under a great navigable river, has long been deemed an important desideratum, both in a civil and military point of view; and two attempts to accomplish it have been made under the Thames. They failed; but the failure arose from causes somewhat foreign to the nature of the undertaking, as is well known to those who are acquainted with all the circumstances. From ignorance of these circumstances however, and knowing nothing beyond the fact of the failure, a considerable degree of prejudice has arisen against a tunnel; so much so, indeed, that it sometimes has been treated even with derision.

To shew the impropriety of such a view of the *present* undertaking, it will be proper to consider it in its different bearings; and also to mention a few circumstances connected with the former attempts; or rather with that of Rotherhithe, for that at Gravesend did not proceed far enough to deserve notice.

An attempt was made in 1809 to excavate a passage under the Thames a little below Rotherhithe, upon a very small scale, and was what, in the language of miners, is called a *driftway*. Its capacity was five-feet high by two-feet nine-inches wide, supported by timber only. No serious difficulty was met with for nearly the whole breadth of the River. They proceeded nine hundred and forty-five feet without any obstacle of importance. Then indeed a considerable body of quicksand came in. This obstruction, however, was soon overcome; and the work proceeded eighty-one feet farther, when it was impeded by a second irruption of sand, within one hundred and thirty feet of the termination of their distance. This second obstruction was surmounted also, and the work was resumed; but the time allowed for the operation being nearly expired; besides which, the ground where it was to commence having been appropriated to the Commercial Dock, and a misunderstanding having arisen among the proprietors, it was determined to abandon the undertaking.

Nothing occurred in this attempt calculated to throw a damp on the present work; quite the contrary; for, even supposing a quicksand were now to be met with, it is clear (from the slender means which proved effectual to stop it then,) that it would immediately be overcome by the powerful means we possess.

Having thus shewn the nature of the former attempt, I will now point out the extent and nature of the present undertaking, from which it will be seen, that there is very little analogy indeed between them. In the present undertaking, instead of an excavation five feet by two feet nine inches, the excavation is thirty-seven feet by twenty-two; no wooden props are used, and a strong brick waterproof arch closely follows the excavation.

In 1823, the formation of the tunnel became an object of deep consideration with Mr. Brunel, the engineer, well known as the inventor of the block machinery at Portsmouth, and many other important works; and his inventive faculty, so ably displayed elsewhere, at last discovered and constructed a machine, where the mechanic powers were so combined as to promise complete success, in the two great objects of supporting the ground, and protecting the men while at work.

He communicated his invention to his friends; and in the beginning of 1824, a number of gentlemen were convened to consider and examine the plans; and all agreed they were not only practicable, but very likely to be crowned with success; this certainly was my opinion, and I became a proprietor accordingly.

It was resolved to form a Company, to carry the same into execution under Mr. Brunel's superintendance. An Act of Parliament was applied for, to incorporate the Company, which was granted without opposition, so satisfied was every one, of the praiseworthy nature of the undertaking. It received the Royal Assent on the 24th of June, 1824, and steps were immediately taken to carry it into effect.

Rotherhithe being considered the best situation, borings were there made in the River, and on its banks, in order to ascertain the nature of the strata; Messrs. Maudsleys and Field, the great iron founders, received from Mr. Brunel instructions and drawings for making the frame or protecting shield of cast iron; a fit and proper situation was fixed on; grounds and houses were purchased, the procuring titles to which, and settling valuations, occupied most part of 1824, and it was only in the Spring of 1825 that the work might be said to have been begun.

. The novelty of the undertaking required much preparation. Before the tunnel could be begun, it was necessary to get down to the depth at which it was to commence, and a shaft or well was formed of one hundred and fifty feet circumference, and three feet thickness of wall; and it was done in a very singular manner; for, instead of excavating to a great depth and laying a foundation, a cylindrical brick tower, forty feet high, was built, of very curious construction.

On the 2d of March, 1825, the Chairman of the Board of Directors, W. Smith, Esq. M.P., accompanied by the other Directors and many scientific gentlemen, laid the foundation stone, embedded in brick, and having a brass plate with an appropriate inscription, coins, &c.

On this occasion, there was an immense assemblage of people, as well in the situations provided for them, as in every place which could command a view. Mr. Smith addressed the meeting, in a way which gave universal satisfaction; indeed, this was to be expected from a gentleman of his high acquirements, extensive knowledge, ready eloquence, and great experience. Amongst other things, he remarked, "That discoveries, equally splendid and useful, are the ornaments of our age.

Out of that steam, which, issuing from a tea-kettle, served only to amuse the eye, or excite a momentary curiosity, chemical science, in its contemplation of causes and effects, has elicited an agent, whose powers, applied in just combination with the mechanical powers of the spindle, loom, mine, and forge, have elevated our country to a rank unprecedented amongst the nations of the globe, by means as laudable as the end is glorious; and proceeding with rapid strides in the path of improvement, has at length superseded the wind for the purposes of navigation, driving the bulky vessel in a direct course across the pathless ocean, regardless of the opposing elements. It has seized the spirit of coal hovering in the mine, and often seen creeping, in a beautiful blue flame, over the rugged surface, and brought it into subjection; on the one hand, restraining its destructive energies with a curtain of iron; and, with the other, compelling it to illuminate our darkness with myriads of brilliant stars, already dispersing their pure light over half the cities of Europe. Nor does chemical science even stop here;—it has analysed the elements, regarded for ages as simple and uncompounded; explained the component parts of the air we breathe, of the water we drink, of the fire which warms us; and, by the combination of substances impalpable and invisible, has exhibited solidity to the touch, and extension to the eye. From these immense progressions of knowledge and power, we derive the highest confidence of success in our bold and novel undertaking. But, when men have done their utmost, it belongs to a superior power to prosper their efforts: to Him we owe the faculties by which we are enabled to discover and to avail ourselves of those mighty agents which He has created; and I trust that this assembly are seriously disposed to join the reverend and respected clergyman of the parish, in imploring the divine blessing on an undertaking in the success of which multitudes are deeply concerned, and the consequences of which may be of such general importance." The ceremony was concluded by the Rev. Mr. Hardwick humbly imploring the blessing of God on this great undertaking, and that it might be for the honor of the kingdom, the happiness of the parish, and the general prosperity of the neighbourhood.

The foundation was laid on a wooden horizontal curb, shod with strong cast iron; and on reaching the top of the tower, at the height of forty feet, there was placed another wooden curb; and the two curbs were connected and fastened together by iron rods passing through the brick work. The ground within was then removed, and this immense structure or tower was found to sink regularly for about thirty-three feet, when it came to a bed of clay, where it stuck fast; thus the *tower* became a *shaft*. The interior of it was further deepened as much as was thought necessary, and it was underpinned (as it is called) for a foundation. The shaft sunk in this manner may be truly said to be the greatest work of the kind ever attempted; and the accomplishment of it, and constructing a reservoir at the bottom, to contain the water which might intrude during the progess of the work, does Mr. Brunel great credit. When all these objects were attained, the shaft was surmounted by a steam engine, and furnished with an apparatus for drawing up and letting down every thing necessary for carrying on the work, which was curiously suspended by iron rods, fixed in large beams, that went across the top of the shaft. The shield, after considerable delay, being at last completed, it was lowered down and placed in the position where the tunnelling operations were to commence; preparatory to which, an opening below in the shaft had been made, of a size sufficient to receive it. This was in the month of December, 1825, when THE TUNNEL may be said to have commenced, for all hitherto done had only been to the tunnel what the coffer dams are to a bridge. The shield may be considered as a moving coffer-dam, destined to secure the excavation in all ways. The difficulty of propelling the shield was very great at first, for want of a proper resisting power to push against; but by the aid of some strong beams abutting against the shaft, it was accomplished.

Mr. Brunel, in directing his attention to the formation of a tunnel on a grand scale, had at once perceived the necessity of completely protecting the workmen, by supporting the ground in all directions, for which purpose the shield, already mentioned, was constructed, and I will attempt to give some account of it; but it is not an easy matter to convey to the mind a correct idea of this machine, without referring to the drawings of it; and, even then, it perhaps is necessary to possess a considerable knowledge of the mechanic powers, before it can be well understood. It consists of twelve frames of strong cast iron, each independent of its neighbour, and altogether weighing upwards of ninety tons; they are three feet wide, and twenty feet high, occupying the whole space from the bottom to the top of the excavation. Each frame is divided into three floors or stories; in each of which a man is placed, to excavate the ground immediately opposed to him; so that they are calculated to contain thirty-six men. All the three men will proceed at nearly the same rate, and their task may be finished at the same time. The frames are then moved forward in alternate order, that is, by advancing every other frame, and securing it in its position; for doing which, screws are attached to them, bearing on the brickwork.

The frames being raised and lowered at pleasure, by screws, press against the top, and support the ground there; and being provided in front with small moveable boards, kept tight by screws pressing them forward, the pressure of the ground in that quarter is resisted, except just at the spot where the workman is cutting. When they have cut away the breadth of one board, they put it up again in its place, screw it tight and remove another, where they again operate until all the ground opposed to their division of the frame is removed; the frames are then moved forward, and the bricklayers build the tunnel close up to them; the men work night and day, by *shifts* of eight hours.

The body of the tunnel is of brick. It is fitted close to the top and sides of the excavation, which is *made good*, as it is called, so that the superincumbent weight is perfectly supported, and the lateral pressure resisted. The bricks are laid in Roman cement, and no *centering* is used for the arch, except a slight iron one, (for the sake of preserving the form of the arch,) which is moved forward with the work. Thus the work of excavation and the building of the tunnel keep pace with each other, and have now (20th April, 1827,) advanced five hundred and forty feet. The soil excavated has hitherto been wheeled in barrows to the shaft, and carried up by the steam engine; but a rail road is now formed below, in the eastern archway, and machinery constructed to bring all the soil out in small waggons, which carry back bricks.

The tunnel consists of a square mass of brickwork thirty-seven feet by twenty-two, containing in it two archways, each of the height of sixteen feet four inches, and of the width of thirteen feet six inches, at the springing of the arch. The carriage ways are nine feet three inches, and the footpaths two feet six inches each. There is a central wall of great strength to separate the two passages, having a succession of arches, some of them so wide that carriages may go from one line of the tunnel to the other. An inverted arch, two feet seven inches thick, is turned under each archway, and supports the external wall and half the middle wall. On this, also, will the Macadamized roads rest. The foundations are all laid on thick and strong beech blanks, which answer extremely well, as there has been no instance of sinking in the foundation, or settlement of any kind. The consumption of bricks is from sixty to seventy thousand a week, and about three hundred and fifty casks of cement; and there are upwards of four hundred men employed upon the works, consisting of miners, excavators, bricklayers and their labourers, carpenters, blacksmiths, &c. &c. for there is both a carpenter's and blacksmith's shop belonging to the tunnel. The length of the tunnel will be about one thousand three hundred feet.

The shaft, with its reservoir, is considerably deeper than the entrance to the tunnel. The excavation for the shaft proceeded through sand and gravel, containing a great deal of water for the first thirty-five feet; this was succeeded by blue clay for seven feet, running into sandy clay, with shells, for about nine feet, which rested upon four feet of hardish stone, under which was coloured sandy gravel, with a good deal of water, for about thirty feet, reaching altogether to the depth of eighty-three feet from the surface. The road-way of the entrance to the tunnel is fifty-nine feet below high water level, or fifty-four feet from the surface of the ground. The excavation has proceeded pretty uniformly through strata which appear to dip about 1½ per cent. from south to north. In examining the strata of the shaft and tunnel, the geologist might find much to interest him, because *so extensive* an excavation in what is called *the London basin of alluvial deposits*, at that depth, has never been seen; and in so far as it is under the

Thames, was, I may venture to say, *never* before to be met with. Between the bottom of the river and the top of the tunnel, there has uniformly been found a stratum of gravel and sand, and then a stratum of strong clay, varying to sandy plastic clay of different colours and consistence, which the miners call *silt*. There next appears this sandy clay mixed with a great abundance of shells, mostly broken. They are of different kinds, generally bivalve, with abundance of oyster shells. It is worthy of remark, that many of the shells are filled with pyrites, or sulphat of iron. Many petrifactions are found, and pieces of wood, with the appearance of being charred, and having their cavities sparkling with pyrites. These specimens of charred wood are, I believe, curious and rare, it not being generally supposed that water as well as fire can char wood. Perhaps the ascertaining this fact may throw some light as to the manner in which coal is formed. Below the mixed stratum of sandy clay and shells, is a stratum of hardish stone, of a calcareous nature, which varies, but may be computed at from two to four feet thick, and it is lost in some places. It is succeeded by gravel mixed with coloured sand. The gravel is evidently water-worn, and generally flint. This must have been deposited when the waves washed their present position, although now from sixty to seventy feet below high-water mark.

It may now be proper to notice the very important part of the business, the expense of the tunnel, and the revenue likely to be derived from it. This naturally engaged Mr. Brunel's attention at a very early period; he made estimates of the expense of every item; and calculated that the whole would not exceed £160,000. But the work has been attended with so many expensive contingencies, occasioning repeated delays, that the calculations as to the progress have not been verified. It was calculated that the tunnel should advance three feet a day; and although sometimes they have gone two and half feet per day, the progress generally has been much slower. If this continues, it is clear, that all those expenses which depend on time, such as wages, lighting, &c. would be increased, and must be added to the above estimated expense. It is true, they may go on better, and with more speed, by and bye: experience may teach a still more expert management of the shield, and a better application of their means; but it is necessary to look at the expense in all its bearings, and not to forget that, in an unexplored road, through an unknown region, there is a *possibility*, at least, that serious obstacles may be met with, although the *probability* of doing so is not great, if we may judge from what we have hitherto seen.

The Act of Parliament allows to be raised £200,000; and an additional £50,000, if necessary, by mortgage. The appropriated shares will only yield £182,000. Let us suppose, however, it may require £200,000 to finish the work; I do not despair of shewing that a revenue may be got from the tunnel capable of paying a fair interest for even that sum; nay, I will go farther, and suppose £250,000 to be wanted; I maintain, that a fair return may be expected for even that amount; at any rate, the tunnel itself will be completed for a much smaller sum; and the descents and approaches will not be *begun* till the tunnel itself is *satisfactorily* accomplished.

Let us suppose, that, by the combination of a variety of scientific and judicious operations, the tunnel, with its descents and approaches, is completed; that all its toll-houses and toll-gates are erected; and that, by a judicious distribution of *gas-lights*, ample illumination is provided: the passenger will find himself placed in a novel situation, where he will have much to admire, and nothing to fear. The gas-light and day-light will blend so gradually at the entrances, that the line of separation will hardly be discernible; and it may be expected, that the convenience of a safe, short, and pleasant passage will be taken advantage of by all classes, and that a *small* toll will be cheerfully paid for so *great* an accommodation. I will endeavour to form some sort of rational conjecture as to the probable amount of those tolls.

The undertaking having the merit of being original, I can have recourse to no case exactly in point. I must therefore, in forming an idea of the revenue, be content with cases that are, in some respects, analogous, such as bridges; and must calculate what proportion the amount of our toll is likely to bear to theirs, considering the comparative population of districts, and the probable number of passengers that may come from a distance.

Experience has decidedly proved that the bridges across the Thames, in and near London, have drawn to them a mass of population, and produced a strong spirit of improvement. Good roads have been formed in different directions, in order to take advantage of the bridges; and thus travellers from the country have contributed to their support. This has been the case not only at Westminster and Blackfriars, but also at the Waterloo and Vauxhall bridges; and it may fairly be stated, that the two latter *were not so much wanted* in their respective situations, as the tunnel is wanted at Rotherhithe.

Looking at the prodigious increase of buildings and population on the North side of the river; in the district of Poplar and Bethnal-green; the numerous great public works, and extensive manufactories, from Blackwall to Wapping, the East and West India and London Docks, the Regent's Canal, connecting that district with the interior;

and considering that the river is a *barrier* which prevents a free communication between the people in the four counties of Middlesex, Essex, Surry, and Kent, on the opposite sides of it, with all the public and private works, wharfs, quays, warehouses, and manufactories; surely, to make a safe way through that barrier must be considered as a grand public improvement, particularly in a severe winter, when, from floating ice, all passing in boats is stopped.

Let us suppose, for a moment, that a fine bridge was thrown across the Thames at Rotherhithe. Can any one doubt, that an immediate intercourse would take place, to a very great extent, between the two sides of the river; that population would draw towards it; that new streets would arise; that the different manufactories on each side of the river would supply each other with the same facility with which they now supply their neighbours, and which, at present, they no more think of, than if they were miles off? Would not travellers from Essex, and the lower part of Middlesex, going to Kent and Surry, and *vice versâ*, prefer this fine bridge and the fine roads, to going round half a dozen miles, and over the stones, through the narrow and crowded streets of London? particularly in winter, when the streets of London are rendered very dangerous, and indeed almost impassable by the accumulation of ice and snow. Would not many of the Deptford and Greenwich stage-coaches, and the Kent and Surry coaches, prefer such a bridge? which would not only facilitate their entering and passing from the city, but would shorten their distance considerably:—there can be no doubt they would thankfully make use of it. Then it is very fair to infer, that, to a considerable extent, the tunnel will supply the place of such a bridge.

A prejudice may exist at first against the tunnel; but as all prejudices, founded in error, give way, when experience has shewn their fallacy, I trust that, in our case, this general rule will hold good; and that almost all the advantages which would result from a bridge, will attach to the tunnel, and that the emolument, which would *to a certainty* be derived from a bridge, will be derived from the tunnel in a very considerable degree.

If a bridge, which would cost almost a million, would yield a competent revenue, how much more reasonable is it to suppose that a tunnel will do so, which will not cost a fourth part of the money.

If a commodious bridge could be built for *what the tunnel will cost*, I presume there are few who would not wish to have a share in it, from a certainty that it would be a profitable concern. But when experience has destroyed all prejudices against the tunnel, why should we not calculate upon its being as much used, and as productive, as a bridge?

Let any one go upon the Waterloo Bridge, and observe the seemingly small number of carriages and passengers; yet, notwithstanding this, the revenue is considerable: he will certainly be a good deal surprised to learn that the net revenue of that bridge is about £14,000 a year. Let him then go to the Vauxhall Bridge, and he will see very few foot-passengers or carriages there; and yet it yields a net revenue of £8,500 a year:— now, comparing the state of the neighbourhood of Vauxhall Bridge, with the site of the tunnel,— where the industry of the greatest port in the world is in full activity,—where hundreds of vessels are built and repaired, and immense numbers loaded, victualled, manned, and provided with all their stores,—all the people connected with these operations living in the neighbourhood; it surely is not unreasonable to conclude that the net revenue, arising from the tunnel, *will equal*, at least, that of Vauxhall Bridge—Nay, it does not seem too much to say, that it will equal the present revenue of the Waterloo Bridge. Let us suppose an extreme case;—that we expend £250,000 which our Act of Parliament enables us to raise; the revenue of the Vauxhall Bridge would give nearly $3\frac{1}{2}$ per cent. and that of the Waterloo Bridge more than $5\frac{1}{2}$ per cent. on that sum.

why the tunnel-shares should be at so great a dis-

works and joint-stock companies. First, speculators, who take shares for the purpose of jobbing, and who sell very soon, if they can realise a profit; if they fail in this, they hold till a call is about to be made, and then they sell, even at a loss, if they cannot do better. Secondly, there are subscribers who look only to the *natural advantages* of the undertaking, to render them a fair rate of interest for the capital employed, and who are willing to wait, till these advantages may reasonably be looked for.

The real value of the shares of every concern depends upon the revenue to arise, in proportion to the capital employed; and the public opinion of such value ought to be influenced by the following considerations. First, the *capability* or *possibility* of yielding *a very large revenue;* next, the *probability* of doing so; and, thirdly, the time that may elapse before the returns commence.

A tunnel or bridge amongst *the Welsh mountains,* for the convenience of passengers, would have no capability of yielding a large revenue—it is impossible it could, the means of an immense traffic not existing in the country; but a tunnel or bridge in the *Port of London* is not only capable of yielding a large revenue, owing to the immense population; but it is even *probable* that such revenue will arise; and that at no very late period after the communication is opened.

The Thames tunnel has created an incredible degree of interest, both in this and in foreign countries. It is the subject of inquiry, to every person who goes abroad from London; and foreigners arriving in London, especially men of science, make a point of visiting the works; nor is it much less the subject of inquiry with persons in England who feel an interest in public and scientific works. I understand there have been amongst the visitors, princes of the blood, many noblemen, both British and foreign, (numbers of our own nobility being proprietors,) his Majesty's ministers, a great many officers both naval and military, and numerous ladies and gentlemen of the first consequence, who have all been to see this wonder of the age.

The Directors, desirous of gratifying public curiosity, as far as it can be done without injury to the works, or interfering with the workmen, have made arrangements to admit the public six days in the week, on paying one shilling; they will be permitted to descend the shaft by a good staircase, and proceed along the west road of the tunnel, for a considerable distance, which even now will place them beyond low water mark, **and**, of course, under the ships navigating the river.*

I hope my fellow proprietors and the public will allow I have redeemed my pledge, that I should remove our *Tunnel Company* out of the pale of *Bubble Companies;* and that they will join me in thinking, complete success promises to crown our undertaking; and although, from the attention of the public not having been directed to the advantages likely to result from the Tunnel, and also from the state of the times, the market price of Tunnel shares may be now at a *low ebb*, yet that I may justly predict they will, ere long, be up to *high-water mark*, and every pound advanced be worth a sovereign.

* I am happy to say, there is every appearance of the above arrangement by the Directors fully answering their expectations: the number of visitors to the tunnel is so great, that a handsome revenue may fairly be calculated upon; and this money the Directors have judiciously determined to set apart as a fund for paying the Proprietors interest on their capital, to increase or diminish with the fund: notice thereof has been given to the Proprietors. An immediate advance in the value of the shares has been the consequence, and will probably continue. We all know what public curiosity is capable of, when a truly wonderful and extraordinary object is presented to it; and I believe I may safely say, that no one who has seen the tunnel ever saw any thing like it, or had the least idea that such a thing was within the reach of human invention. It is but fair, therefore, to conclude, that the revenue will very considerably increase, and that a handsome sum, as interest on capital, will be divided amongst the Proprietors.

THE END.

IMPORTANT BOOKS,

Published by Effingham Wilson, 88, *Royal Exchange.*

STATISTICS OF THE BRITISH EMPIRE;

In 1 vol. 8vo. price 15s.

Statistical Illustrations

Of the TERRITORIAL EXTENT and POPULATION, RENTAL, TAXATION, FINANCES, CONSUMPTION, COMMERCE, INSOLVENCY, PAUPERISM, and CRIME, of the BRITISH EMPIRE; arranged and published by order of the London Statistical Society, from upwards of one hundred folio Volumes of Journals, Reports, and Papers, laid before Parliament during the last thirty years.

In 8vo. price 5s. boards,

A NEW CHECK JOURNAL,

UPON THE PRINCIPLE OF DOUBLE-ENTRY;

Which exhibits a continued, systematic, and self-verifying Record of Accounts of Individuals and Partnership Concerns, and shows, at one view, the real state of a Merchant's or Trader's Affairs, by a single book only, even should a Ledger not have been kept; whereby the tedious repetitions in journalizing (inseparable from the present practice) are wholly avoided, the Balancing of Books made a speedy and easy operation, and the use of the common Journal and Cash Book rendered unnecessary; combining the advantages of the Day-Book, Journal, and Cash-Book, and by which a saving of three-fourths of labour is obtained; with particular Forms for Merchants, Under-writers, Wholesale and Retail Dealers. The whole familiarly explained, and forming a complete and practical System of Bookkeeping by Double Entry, the result of many year's experience. Wherein Interest, Joint Adventures, and Joint Purchases are particularly treated upon.

DOUBLE ENTRY.—"Unless a shorter and more simple method is devised for obtaining the same results, it will continue to deserve the preference of the Commercial World."—*Cronhelm's Remarks on Double Entry.*

"Book keeping, as it respects genuine practice, is at a very low ebb indeed, and consequently needs improvement."—*Jones's Observations on Balancing Books.*

By GEORGE JACKSON.

Also by the same Author, Second Edition, corrected, enlarged, and greatly improved, with a coloured Frontispiece, price 1s.

POPULAR ERRORS IN ENGLISH GRAMMAR,

Particularly in Pronunciation; familiarly pointed out for the use of those Persons who want either opportunity or inclination to study the science.

ANDERSON'S LONDON COMMERCIAL DICTIONARY
And General Sea-Port Gazetteer,

With the Consolidated Duties of Customs and Excise brought down to the present time; a List of all articles of Commerce in every European Language, with their marks of excellence; the Weights, Measures, Monies, Commercial Laws, and Regulations of all Nations. A new Edition. In one very large 8vo. volume, price One Guinea.

Stereotyped Edition of BOOTH'S INTEREST TABLES, free from Error, in 4to. price £1. 16s. boards, or strongly bound in calf, £2. 2s.

TABLES of INTEREST, ON A NEW PLAN,

By which the Interest of any Sum, from One Pound to a Thousand, from One to Three Hundred and Sixty-five days, will be found at one View, without the trouble or risk of Additions. Also, the Fractional Parts of a Pound and from One to Ten Thousand Pounds, at Five per Cent. To which is added, a separate SUPPLEMENT, that renders these Tables equally applicable to any other Rate per Cent.

By DAVID BOOTH.

"I have examined 'Mr. Booth's Interest Tables;' the arrangement is novel and perspicuous; and I have no hesitation in affirming, that the Work will be far more useful to the Public than any one which has hitherto appeared on the subject. "CHARLES CARTWRIGHT, *Accountant-General to the East India Company.*"

In One Pocket Volume, price 5s. 6d. boards,
THE TRAVELLER'S & MERCHANT'S FINANCIAL GUIDE
In France and Flanders;

Containing Tables, reducing Sterling Money into French Currency; with a Scale of the Weights and Measures of France, with the equalised Proportion of those of England.

By JOHN NETTLESHIP.

Lightning Source UK Ltd.
Milton Keynes UK
UKHW012247051118
331648UK00013BB/2059/P